YOUR KNOWLEDGE HAS VALUE

- We will publish your bachelor's and
  master's thesis, essays and papers

- Your own eBook and book -
  sold worldwide in all relevant shops

- Earn money with each sale

Upload your text at www.GRIN.com
and publish for free

Bibliographic information published by the German National Library:

The German National Library lists this publication in the National Bibliography;
detailed bibliographic data are available on the Internet at http://dnb.dnb.de .

Imprint:

Copyright © 2013 GRIN Verlag, Open Publishing GmbH
Print and binding: Books on Demand GmbH, Norderstedt Germany
ISBN: 9783668290563

This book at GRIN:

http://www.grin.com/en/e-book/339024/analysing-project-management-practices-
in-ofc-laying

Mudasir Dar

Analysing project management practices in OFC laying

Critical areas and challenges delaying OFC laying for a government monitored project in Kashmir

GRIN Publishing

Analysing project management practices in OFC laying

Critical areas & challenges delaying OFC laying for a government monitored project in toughest terrain of Kashmir

Mudasir Dar CSSBB® PMP®

Abstract

In OFC laying, execution approach for risk management using set of practices are normally insufficient which limit the success of on time project delivery. This paper describes the approach for managing execution and provides set of practices for completing an OFC laying project on time within budget.

In developing countries like India, project management for developing OFC infrastructure is still not mature and there is lack of coordination between various OFC deployment agencies. This paper provides insight for proper planning of OFC deployment and analysing the case study of project delay in a critical OFC deployment project.

The study concluded that there is a good opportunity for implementing best project management and improving execution practices, for deploying OFC in the toughest terrain of India, where all kind of challenges are encountered.

Introduction

Project management as we know it today, or conventional project management, emerged in the 1950s in the defense and aerospace sectors. These sectors in this timeframe can be characterized as little flexible and complex (Morris, 1997). Starting in the 1990s and still growing is the awareness of the changing and dynamic project environment (Bosch-Rekveldt, 2011). It is recognized that the complex and changing context of a project makes it impossible to make reliable predictions, and instead of predicting and correspondingly avoiding changes, changes need to be incorporated in the project (Priemus, Bosch-Rekveldt & Giezen, 2013).

Project management for OFC laying involves coordination between numerous agencies particularly in a government funded project. Planning, risk mitigation, documentation, contract management and execution strategies are very critical for project delivery and needs to be prepared beforehand with proper brainstorming, with various subject matter experts and selected people from the local region.

OFC laying is a complicated project wherein every end user need the services without having any inconvenience during the OFC deployment. OFC laying generally involves following steps

1. Route survey and approval by customer.
2. Identification of various right of way authorities.
3. Allocation of sub con and awarding contract to carry out the work.
4. Application for right of way and subsequent payment of right of way charges for getting formal permission.
5. Planning for resource deployment and mobilisation of resources to the site.

6. Trenching and ducting using different type of machines like HDD, JCB, and rock breaker etc.
7. Installation of Manholes and Hand-holes at proper locations depending upon customer requirement.
8. Duct integrity/Duct clearance test to ensure duct is free from kinks and obstructions.
9. Cable blowing through the duct.
10. Splicing and termination of open ends of cable.
11. Link testing and rectification of losses at joints and other bends.
12. Link commissioning to customer with proper documentation.

Studies show that causes of poor performance of a project can be divided into external causes and internal causes (Meng, 2012). External causes, which are usually beyond the control of project teams, may include adverse weather conditions, unforeseen site conditions, market fluctuation, and regularly changes while internal causes of poor performance may be generated by the client, the designer, the contractor, the consultant and various suppliers who provide labour, materials and equipment (Assaf & Al-Hejji, 2006).

Project delivery has recently started to require a flexible approach. This need results from the fact that projects are delivered in a changing, and therefore more unpredictable, environment, which translates directly to the way a project is conducted and the interested parties' expectations towards its outcome. Such a situation makes it complicated to plan the course a project might take as well as subsequently deliver the project according to the accepted flow.

Project management methodology guidelines can play a critical role in reaching to project success of a complex project like OFC laying and needs to be tailored in certain conditions. PMBOK® Guide (2013), model of project delivery is quoted. It consists of the following stages:

- Initiation (defining);
- Planning;
- Performance (the actual delivery);
- Progress monitoring and control;
- Closing the project.

Kashmir is altogether a different challenge, wherein, on an average 6-10 days per month are lost in various disturbances strikes, curfew and rainfall. In the current case study few more challenges like laying OFC in areas where insurgency activities were a normal routines were encountered. Terrain in higher altitudes is very hard wherein specified depth cannot be achieved and even in some places there are non-motorable routes posing a hurdle to laying of OFC in such areas. Softer soil is mainly found in lower regions of Srinagar and Anantnag districts and few areas of Baramulla where HDD methodology can be implemented. Regions like Uri, Kupwara, Bandipura, Dawar purely consist of hard rock where HDD methodology is hardly impossible.

	Average Percentage distribution	
	HDD	OT
Srinagar	75%	25%
Anantnag	67%	33%
Kupwara	15%	85%
Baramulla	54%	46%
Bandipura	20%	80%
Uri	5%	95%

Table 1. Distribution of HDD&OT methodology based on survey in different districts of Kashmir

Executing an OFC laying project for a government funded organisation is itself a complex situation which requires specific actions to avoid delay and failures. The purpose of this study is to explore all challenges in OFC laying and possible solutions to complete project on time within budget.

Discussion

In OFC laying for a region with multiple challenges, project management knowledge is a must. Critical areas for focus to deliver the OFC project in a region with all kind of challenges can be categorised as

1. Planning.
2. Resource management.
3. Contract management.
4. Human resource management.
5. Risk management.
6. Documentation management
7. Execution monitoring and control.
8. Strategy management.
9. Customer relationship management.
10. Closing of Project.

In this case study, all focus areas mentioned and challenges in each area will be analysed. Possible solutions to each challenge are discussed and have been proposed for implementation.

a. Planning
a.1 Pre-bid planning
It's a normal practice in India to plan things in a very limited group of people to keep secrecy of project. The pressure from top management to get more business sometimes skips some key aspects in planning while bidding a government funded project. In current case study, planning was done without involving expert people from local region to know about challenges and preparing a risk management plan

according to the ground conditions. Key factors impacting project delivery plan were not considered and such impacted project delivery. Current project required deployment of OFC network for more than 9000km in a limited span of 18 months with an altogether different challenges in a most disturbed area of India.

It's evident that we cannot use general calculations for project delivery in scenario which is completely different. There have been general macro-models used for calculating timeline of project in OFC like

Time for project delivery={ Scope of work (in 'km')/{(No of machines which can be deployed * Average output from one machines in 'km'} + Contingency days.

e.g for 9000km of execution with 100 resources(JCB/HDD)/day, and an average output of 0.15km/day with 100 contingency days,700 days(~2 years) are required to complete the trenching& ducting of project.

Usually trenching and ducting constitute 40% of weightage for project completion, therefore additional 1050 days will be required for other activities like right of way, cable blowing, splicing & termination and link testing/ commissioning. Thus for executing 9000km of OFC in a normal region, around 4.8 years are required.

Even for above example if resources are doubled to 200, around 2.8 years are required. Such kind of calculations doesn't hold well for a region like Kashmir and other similar regions. There are regions in Kashmir where only 6 months are available for execution and are under snow for rest of the year with more than 8-10 feet of snow. Such regions are even very difficult to execute where sheet rock mountains itself create a challenge for digging. While planning of the project this part was not considered properly which led to delay project delay. This is only an example and may or may not be part of every OFC laying project.

While planning for OFC bidding, for the sake of development of country, a common meeting must be held between key telecom operators to know about their OFC deployment plan and lay additional duct wherever there is OFC requirement. This will ensure that money is not wasted in digging trenches for OFC laying again and again. Also OFC highways should be planned in a country like a India which is having developing OFC infrastructure.

Solution to pre-bidding problem

While bidding for OFC laying project, local expert's feedback, topography, political situation and cultural variance must be studied properly before rushing for bidding for new business. Local expert opinion must involve all regions of the particular location. All risks which are pointed at pre bidding phase should be taken seriously and mitigation plan must be prepared in advance. Normally risks are listed during pre-bidding but mitigation plans are never reviewed with a focussed and separate agenda. Proper review mechanism should be in place for only risk management.

A practice of finalisation of bid document before one week of last date of bidding should be inculcated in organisation so that no important aspect is skipped during rush for last date and management pressure.

a.2 Resource planning

In a complex project like OFC laying where work can get hampered even sufficient resources are deployed on ground, resource planning is indispensable for timely completion of project. In the current case study the exact scenario of resource availability was not analysed and project was initiated. In a time-limited OFC laying project such kind of delays have a cascading effect and the time loss gets multiplied at each stage.

Sourcing of resources were not analysed by the concerned stakeholders at initial stage and such the problem was not put on table thereby no focussed action taken towards making resource available. Even in most of the management reviews resource requirement discussion was a routine but no focus and systematic approach was followed. It's a normal practice that SCM team in order to hide inefficiencies, link resource with non-availability of work front. Never ever a team should be over confident in terms of making resource available.

After procurement of resources, planning should be done in such a way that idling time for existing resource is negligible. In current case the resources available were not fully utilised as different support functions try to pass the ball to one another. Right of way responsible person points finger towards ASP issues like liasoning and trench backfilling issue. In OFC laying ASP has to liason with various ROW authorities to obtain formal permission and due to delay in ASP response the entire chain gets effected. On other hand after execution its responsibility of ASP to backfill the excavated material in proper manner so that it doesn't cause in-convenience to public and concerned right of way team.

Solution to resource planning

Resource planning after project initiation has to be clear and transparent and any problem that can be pointed at initiation stage must be taken in positive manner and incorporated in risk register. Resource and supply chain team must have proper forecast of resource source and must be clearly articulate to management. Any gap in resource requirement has to be acted upon from initial stage of project rather than realising need during project execution. Key factor for success are

1. Whatever resource is required for completing project within time, 5 times resource source need to be identified in case of a region like Kashmir where most of suppliers are not interested. All information like mobilisation time, expected run rates, cost of rentals must be recorded at beginning of project e.g if rental team is to be hired what will be cost, where is the source, what is mobilisation time, number of sources and formal quotation from all concerned. Focussed team for addressing this key challenge should work only and only for this matter with no other responsibilities. At no point of time this team should be dissolved during course of project.
2. Project execution, SCM and ROW team should formally review each other at regular intervals with specific agenda and issues related to resource idling and deployment. In normal cases teams sit together but without specific agendas and meeting are very informal. Top management should review action points from such meetings in their

governance framework. Each function should be made accountable for project delivery equally.

3. A backup backfilling team who should work under support function of right of way team to address any issue not addressed by execution team, to avoid any resentment of right of way authorities and public. This will stop the blame game to a greater extent and will speed up right of way permissions.
4. Miscellaneous cost of each OFC link should be tabulated, which must include idling cost of resources, cost of visit to the link, cost of backfilling, and other costs which relate to damages. This has to be reviewed by finance & accounting team to have proper understanding of loss of resources and same need to be made available to owner of links in a predefined time.
5. Resource diversion route map should be prepared in advance, in case of work stopped by authorities or public.
6. Resource manager must be made accountable for proper utilisation of resources.
7. Resource requirement must follow a smooth inclined curve rather than parabolic curve.

Implementing sound organizational project management governance framework can enable the kind of visibility and control that are essential to successfully deliver the expected benefits from resource planning.

b. Resource Management

During resource planning macro factors are being studied while as in resource management micro details are analysed. Resource management is the heart of OFC laying project and resource planning is the aorta of the heart. Road maps of resource procurement, resource deployment, resource diversion, resource ASP, resource risks, etc must be prepared and followed.

Resource manager should be empowered to take decision for movement, deployment and removal from sites without any intervention from sourcing team. All commercials for such movements must be decided before finalising contracts. Costs involved in movement should be linked to costing budget of the links.

Resource manager should analyse the availability of resources on rentals and involve some fund for procurement of resources which can be used, moved and deployed at ease without any interferences from sub cons. One of the biggest issue while making resource available is that sub cons have resources available but are occupied in other work as well. While estimating the availability date, all such information must be considered.

Proper tracking mechanism should be installed on procured machines to avoid any mischief by sub cons who operate them.

Also for huge rollouts of OFC, there must be tie ups with different organisations for supply of labours and machines to reduce loss of time in mobilisation and finalisation of contracts. Even in last minute requirement these tie ups will facilitate resource deployment without losing time.

Also it is suggested if an organisation who is a pure OFC rollout company, must develop in house organisation for labours and machine supply. These initiative will help the project delivery in totality, besides revenue generation from supplying resources to other telecom operators.

Contract Management

In current study, it was observed that there was big flaw in contracts which was one of the major causes for delay. Work allotted to different sub- contractors was done randomly due to lesser number of available sub-cons in region of Kashmir and dis-interest of outside sub-cons to work in Kashmir. This resulted in a selection of ASP who are hardly aware about the execution of OFC work and who started work with lot of experimentation.

Since during initial phase of project, ASP finalisation process was very slow and as soon as it came closer to execution start up, negotiation were done seriously and few ASP were taken on board with higher prices. The inter lock breakup through on boarding of few vendors resulted in on boarding of other ASP in Kashmir. In this crowd all types of ASP were taken on-board irrespective of their capability. Documents produced as a part of eligibility were not verified properly.

During the allotment of work, no delivery timelines were signed off with ASP which allowed ASP to work as per their convenience and comfort. No penalty clauses for non-standard work were incorporated in contract. Due to non-standard work rework was done which delayed the project.

Vendor governance is a very critical for understanding both side issues and it helps to keep system transparent and on track. This ensures that all measures are taken from both sides for on time project delivery.

Flaws that were pointed in contract were not even rectified in due course of time. Later half of the project kick off with ASP was started but the system to monitor the contract was inappropriate.

Solution to contract management issue

1. Delivery timelines must be mentioned in contract and if right of way issue is problem of policy then hindrance for same has to be informed and recorded. Action owner for each hindrance must be made accountable and accordingly target should be set. In this case focussed review mechanism for resolution of such issues should be discussed in governance meeting.
2. One of the best solution is to fix rates as per delivery timelines **(Variable rate model)** e.g. for a particular job if minimum(best possible) time required for completion is 30 days, then rates should be fixed accordingly.
   Rate (model for above example) =Max{[(Best possible time for completion/Time elapsed by contractor for completing job) * Max rate that can be offered for the particular job.], (Minimum rate for the job as per expenditure)}

3. Contract review must be done fortnightly with all ASP and dashboard for each ASP should be floated on notice boards. Contractors who are not performing should be eliminated from system with no delays after giving formal warnings.
4. Penalties for non-standard work must be mentioned in contract so that ASP is aware for not doing things First Time Right. One of the key success for completing OFC laying project is first time right approach.
5. Vendor governance must be conducted regularly in a structured manner wherein both side issues must be discussed with clear ownership and target dates should be signed off.
6. Automation for data display, for various contractors wherein various deficiencies should be pointed with help of data and corresponding feedback should be given to them for their improvement.

Human resource management

One of the key success for setting up new business is hiring right people for the right job. It's a time consuming process with monetary losses if person hired is not fit for the job. In a new project especially project being undertaken by a matrix/functional organisation, it's a challenge to choose right people for right job. Hiring people under pressure from execution and management team is a serious blow for the future of project.

In current study same issue could be observed as people who have ONLY worked in telecom were preferred to be taken on board randomly due to hiring pressure. In OFC laying technical skills are least required and people with passion, temperament, excellent communication skills are required most. People will hardly need training for few weeks to train them for OFC laying.

The kind of passionate people required to execute OFC and that too for a government funded project particularly in Kashmir, could not be made available at the right time. People were hired with excessive hikes but right people with passion were missing.

People hired, joined organisation and left organisation within span of 3-5 months. This was a setback for the project execution and lot of documentation was missing when people left organisation. Reason for leaving organisation can be categorised as

a. Immense pressure to complete un-realistic targets without ASP support.
b. People not accustomed to such kind of requirement like run rates, documentation etc.
c. People not fit for the job
d. Security, political and climatic issues in Kashmir.
e. In genuine demands from government monitoring agency, people had to move door to door for a small signature created frustration.
f. Cultural fitment and high cost of living in Kashmir.
g. Cohesive relation between manager and his reportee could not be setup as output desired from day one with immense pressure.

Lot of people retired from government organisation were taken on board was also major flaw for the project execution. Working style of people who have worked in Govt. organisation cannot set pace with corporate world. Excessive ego and fear from seniors exist in people from government organisation which created a gap in delivery and demand and handling government agencies.

Excess hiring of people also allowed many people to move freely which led to inefficiencies in the project. More than 30% people hired were not accountable for quantifiable output. People who had very less work tried to peep into another function and also made them non accountable toward their own key responsibilities, which was not good for the health of project.

There was a complete imbalance in fulcrum of organisation. Few people were extremely loaded and few people were having no work.

Solution to Human Resource management

1. Getting passionate people on board through face to face interview rather than telephonic only.
2. Proactively interview should be conducted and people must be kept in waiting list for any future vacancy to avoid hiring pressure.
3. Training people for OFC laying and stress management.
4. Empowerment of HR team for reviewing people performance and issues faced by people on fortnightly basis.
5. Signing of agreement and bond for retaining people for a minimum period.
6. Retention incentives should be proposed for prospective candidates.
7. Engaging people in different activities like sports, get together's etc.

Integrated team responsible for all activities whether ROW, execution, quality and handover
Risk Management

Major risks that can impact project badly were either ignored or not acted timely. Key risks like ASP availability, climatic challenges, and work day's availability where mitigation could have been devised in advance were not acted timely. This led to cascading of risk impact on the project delivery timelines. In an OFC laying project most of the challenges fall in external category and if any work is delayed due to internal issues, this can lead such kind of projects nowhere have greater repercussions.

Various issues of ROW policy were encountered which remained unresolved for 90% of time of project due to lack of urgency.

In an OFC project following risks must be looked upon

1. ASP availability in a particular region.
2. ROW policies of various agencies.
3. Resource availability in region.
4. Fund involvement and cash flow.
5. Snow laden routes working window.
6. Availability of protection materials for OFC laying.
7. Unrest and climatic challenges.

8. Customer relationship.

Solution to risk management

1. Maintenance of risk register for scheduled formal reviews
2. Identification of all risks during pre-bid and execution of the project. It should be incorporated as a process in organisation.
3. Focussed action on risks rather than discussion only during reviews.
4. Brainstorming with execution on future risks of the project.
5. Policy/process to be made for keeping healthy customer relations.

Documentation management

In a project, monitored by government agency, documentation management is a critical success factor. In the current case study, clarity for documentation was not clear from day one. Most of the people who joined the project have done projects in private organisation where documentation was not that important. Once people joined this project, during 70% of project time, documentation made was either incorrect or not made. There was capability issues as well with the team taken on board.

Documentation which was supposed to be made daily by engineer on site was never done or followed. This is what we call team leader failure; to check this important aspect of project, when it was clearly articulated in a span of 1 year. People pretend to be busy in non-value added works and ignoring the documentation part of the project. Documentation review team was also not in full action to control documentation of the project. This was a big hit to the project as people energy wasted in making documents again and again.

After proper documentation another challenge for the team was to get it signoff from customer and government monitoring agency. This again led to resource utilisation and engineer spending most of the time in office rather than monitoring execution on site. This problem was due to three reasons

a. Documents were not prepared on site regularly by site engineer.
b. Customer relationship team not clearly communicating customer and monitoring agency to signoff documents on site which was an agreed process and norm.
c. Clear guidelines for documentation preparation

If we analyse this problem, it's very easy to make a single page document on site daily. An engineer will hardly take 15minutes in a day to make documentation for his allotted site in a day. This is a habit which should have been inculcated in team by their team respective team leaders.

Solution to documentation management

1. On site daily document preparation.

2. Fortnightly document review on sites and dashboard to be prepared by document controller
3. Document controller position to be created in organisation.
4. Strong customer relationship team who can defend any proposed document change during project execution. Every change must follow change management process.
5. Incentives to team for 100% document readiness for a fixed period of time.
6. Training for document preparation on site, with scheduled review for success of training.
7. Documentation automation. Daily updation of progress through software tool/app which will create further documentation automatically.

Execution monitoring and control

In an OFC laying project, execution monitoring on site, is one of most critical success factor. Site visit must be done regularly by engineer and at least once in a week by respective manager. Without site visits, an OFC laying project will never complete qualitatively. Execution monitoring will ensure that work done is within quality specification and any issue on site is not hidden from concerned stakeholders. Many issues on site can be easily resolved if focussed and site visit being conducted regularly.

In current case study, 50% people who are responsible for daily site visit hardly visit site once in a week. They report the happening of site by taking information for contractor supervisor from their residence and update accordingly. Even site photos are being taken in a similar way to report to superiors.

Benefits of site visits are

a. Proper layout of route where all hidden information can be earthed out. Such information can be utilised in advance and problems which can be encountered during execution can be mitigated in advance.
b. ASP working on site always tend to do shortcuts which will continue if not monitored.
c. Better relationship between ASP/contractor and engineer/manager.
d. Better planning can be done if entire route visit is being done regularly.
e. Work can be inspected and certified on daily basis to avoid backlog for inspection.
f. Monitoring quality of backfilling which create major hindrance for continuing work falling under different ROW authorities.
g. Ensuring ASP fills the excavated trench material manually before using machines. This problem creates biggest challenge for blowing OFC through the PLB duct.

Trenching work usually involves two method

a. Open trenching(OT)
b. Horizontal directional drilling(HDD)

HDD is faster mode of trenching and speed is usually 200-500% faster than OT. HDD is done using a machine which can give output for 100m to 1000m in 8 hr span of day depending upon strata and topography. This type of methodology is cheaper and creates minimum inconvenience to public and ROW authorities. HDD methodology is easily allowed by ROW authorities and ROW charges are minimal. This methodology is usually applicable for softer soil. HDD are also available for harder strata but cost of operation for such machines is very high and in such cases OT is preferred.

OT method of execution involve lot of efforts in obtaining permission. This method is slower than HDD. Public inconvenience and damage to road assets is key challenge for adopting this methodology. Lot of liasoning is involved in getting ROW permissions. OT is usually done by JCB or manually by labours. JCB executes 50-200m /8hr and one labour execute 1-3m/8hr in a day, depending upon strata. Whenever any hard patch is encountered rock breakers are used to get specified depth.

Various execution challenges that are faced are

1. Trenching speed very low.
2. Frequent work stoppage by various agencies.
3. Issue with backfilling with soft soil to avoid kinks in duct.
4. De-coilers and end plugs not being used religiously.
5. Coupling between two ducts, not done properly, resulting in air gaps.
6. Public issues where new road is constructed.
7. Policy issues of various ROW agencies.
8. Regular commitment failures from ASP.
9. Workmanship issues resulting is rework.
10. Procedures for DIT and blowing not adhered.
11. Deployment of required resources on ground by respective ASP.
12. Customer priority.
13. Rainfall, snowfall and security issues.
14. Managing resource haltage during work stoppage, to avoid ramp up after work resumption.
15. Installation of Manholes and hand holes
16. Patch work done by ASP.

Solution to execution monitoring and control

Trenching

1. Before starting execution of any OFC link, recorded video and quality person certification should be obtained to ensure OSP is equipped with required accessories like decoiler, Champhering tool, duct cutter, safety accessories etc. A proper signoff must happen between ASP, quality and project to ensure same will be used during entire execution and if found guilty agreed penalties must be agreed.
2. Detailed route survey should be done to know route exactly and devise mitigation methodology in advance. Sample pits must be done on entire route to check the strata and convert permission from HDD to OT if required. This problem lead to resource haltage if not addressed proactively. Sample pit report should be submitted and recorded.
3. Resource deployment points should be discussed and agreed with ROW agencies. Key success factor is to gain confidence of ROW authorities by proper backfilling up to their

satisfaction. Even in some cases monthly meeting should be done along with concerned engineer. Specific action points should be taken in a positive manner and resolved through concerned stake holder within agreed timelines. An Alternate ASP should be aligned if man ASP takes time to resolve backfilling issue.

4. All protection material and rock breakers must be made available on site as per sample pit report.
5. Any difficult patch or patch with local hindrances should be discussed in reviews and focus should not get away from it till it is resolved.
6. Plan of execution should be signed off with customer to avoid last minute changes.
7. In case of OT work manual backfilling (upto 30cm) to be monitored by engineer and certified to avoid kinks. Soft soil if not available on site must be purchased and mentioned in ASP contract to avoid arguments for such cases.
8. Each coupling to be certified by an engineer to avoid time loss during blowing and same should be recorded.
9. For manhole and hand hole installation an alternate vendor should be identified and deployed if ASP doesn't respond. Double the cost should be debited from vendor for allowing us to deploy alternate vendor and same has to be mentioned in his contract. This is very important as main ASP normally completes trenching and ducting work and claims same for billing. Also billing must be kept at blowing rather than trenching and ducting completion.
10. JCB can be deployed in a span of 2km provided backfilling is done properly in same day. This will increase T&D pace without causing inconvenience to public and authorities. Resource deployment should be planned in such a way to achieve target for days lost in rain, strikes etc along with current month/week target.

In current case study engineer spend most of time in other activities rather than focussing on his key responsibilities. This is a big blow for first time right approach. For any ASP related issue, SCM team must be involved in a day to day basis so that engineer focusses on his key responsibilities rather than chasing vendor and customer signoff.

DIT, Blowing & Splicing

1. Before starting blowing, front creation process for blowing must be started so that as soon as blowing team reaches site only OFC will be blown rather than opening of pits and duct clearance. During OFC blowing lot of time is wasted if duct clearance is done side by side.
2. Proper DIT procedures must be followed and certified by engineer and kept in records.
3. Gap between cable outer diameter and inner diameter of should be uniform while maintaining air pressure inside duct.
4. Automatic/hydraulic jacks should be provided on site to reduce time loss for unwinding cable.
5. Proper lubricants should be used to reduce coefficient of friction between OFC and HDPE duct.

6. Air Compressor of proper range should be used for blowing cable and DIT which have after cooler mechanism.
7. Splicing must be done in closed environment.
8. Cleavers must be checked regularly to check for their efficiency.

Strategy, customer relationship management

If a matrix/functional organisation take a massive project of OFC laying first time, strategy of organisation should be aligned with projectised organisation. Functional organisation have different set of mind-set and there are changes required in strategy to forecast the expected problems during execution.

In current case study various strategies like execution being done directly through ASP, through strategic partners etc. were implemented. Rigorous homework that was required for selecting strategic partners was not done properly due to limited time window which resulted in termination of contracts, huge loss of time and organisation reputation, when strategic partners were not performing. It's better to get execution directly through ASP as it saves cost and have better control on inputs to deliver outputs. To crash the project, strategic partners should be involved only after proper evaluation.

The kind of changes done in organisation structure frequently showed kind of immaturity in projects. Changing organisation structure within same set of people only incur delay, in a strict time bound project. This also lead to conflicts in team. If non -performance is criteria to change organisation structure then people with different mind-set should be incorporated and small changes should be implemented. This also lead to conflicts in team. Strategy of hiring people from customer side **only** in relationship management department was also not a good decision and only overburdened cost centre as required objective could not be met. It's good to have one or two people in Customer relationship management from customer organisation but not driving entire department through them. There must be mix of people handling customer to ensure corporate culture and urgency to deliver, remain prevalent across life cycle of project. The kind of urgency and sensitivity in people who have served most of the time in govt. organisation is not matching the expectation of corporate world.

Customer handling needs specialised and polished skills to get the right work done. Request and demand must go hand in hand which can only be achieved in mix culture. Also the importance and priority of project should be articulated in customer organisation clearly and concrete steps must be taken to ensure that priority is communicated to the entire chain.

People who are responsible for cost centre and HR in organisation should review leadership team to understand actual causes of non-performance and increased cost. Increased cost will give you an idea where the actual problem lies and where the hammer has to be hit to secure the project. People from projectised organisation culture should be hired to spread the culture of project management.

Journey from functional to projectised organisation is indeed a tough phase where road map should be clearly drawn from top management. This process can be smoothened by regular consultations and benchmarking with projectised organisation.

Closing of project

In a govt. driven project, documentation is very critical for closing of project. All stakeholder must be informed formally about completion of project and proper handover document should be kept in record. All documentation like As-built diagram should be handed over to operation and maintenance team to ensure health of the link is maintained during life cycle of network.

Conclusion

In an OFC laying project each individual should know his key responsibilities and deliver as per that. Frequent site visits and regular review of documentation should be done. Only required issues should be reviewed rather than diverting the focus of review meetings.

Planning deployment of OFC with consideration of all constraints, resource management through structured manner keeping a diversion plan in place, monitoring and control to be done only from actual work sites, vendor governance and contract management must be aligned with the project objective/delivery timelines, and extensive use of technology in automation of document control are very critical, which needs a focussed approach throughout project life cycle.

Also creation of in-house organisation for supplying labours and machinery for huge rollouts, which will minimise dependency on other suppliers is recommended for reducing delay in resource mobilisation.

Key recommendations from study should be implemented in project scenario like this and learnings of this project to be taken forward.

References

1. Meng, X. (2012). The effect of relationship management on project performance in construction. International Journal of Project Management,30(2), 188-198.
2. Assaf, S. A., & Al-Hejji, S. (2006). Causes of delay in large construction projects.
3. Morris, P. W. (1997). *The management of projects*: Thomas Telford
4. Bosch-Rekveldt, M. (2011). *Managing project complexity: A study into adapting early project phases to improve project performance in large engineering projects.* Delft University of Technology.
5. Priemus, H., & van Wee, B. (2013). *International Handbook on Mega-projects*: Edward Elgar Publishing.